# A VERY Brief Guide to AQA GCSE Physics 1

**A revision guide for those in a hurry.**

First published December 2015
ISBN: 1519632347
ISBN-13: 978-1519632340

# Foreword

AQA GCSE Physics is a popular exam course. It shares two units (P1 and P2) with the GCSE Science course. This book provides a quick look at the content of Unit P1.

The author has an MA in Natural Science and an MSci in Experimental and Theoretical Physics from the University of Cambridge. He has taught physics in state and independent schools for over 15 years. He is a lead examiner of physics A-level and an examiner of physics GCSE.

# Introduction

## About This Guide

This revision guide is split into topics in line with the AQA specification. All the key points are covered in a clear bullet point format.

## Revision Technique

Effective revision is that which starts early and is frequently added to. Whenever you finish a topic in lessons, you should revise it. Every month or so, you should take a look back at earlier topics and refresh your memory.

Effective revision is active. Reading through text is a good start, but you are unlikely to remember everything. Instead of passively reading, you should write things down, for instance on cards or on a poster. One of the best things to do is attempting questions – you can download past papers from the AQA website (http://www.aqa.org.uk/subjects/science/gcse/physics-4403/past-papers-and-mark-schemes) or buy my Physics Problems for GCSE book (http://www.archaeoroutes.co.uk/edphys/problems.php).

Using This Revision Guide

This book lends itself particularly well to use throughout the course. However, it is also eminently suitable to use for revision in the run up to the exam.

Here are a few ideas of ways you can use it:

1. Read a topic and write it out in your own words.
2. Get together with friends, all read the same topic and discuss it.
3. Work through a topic and annotate or correct your class notes with things you missed or wrote down wrong.
4. After a lesson, find the appropriate section in the guide and check that you have made complete notes.
5. Look up anything you don't know, for instance when doing homework.
6. Get a friend or family member to invent questions and then test you. After a while swap over – writing questions is as good a revision technique as answering them.
7. Go through a topic and traffic light it. This means colouring each bullet point green (if you are happy), yellow (if you are a bit unsure) or red (if you have no idea) so you can target your revision where you need it most. *If you are reading this as an ebook, your reader probably has highlighting tools you could use.*
8. Teach a lesson on a topic to a friend or family member. They should have the guide in front of them to tick things off as you go through them. They can then give you feedback.

Another really good use of this revision guide is as a focus for tutorials. You could work through a topic and then ask your tutor/teacher about anything that you didn't understand. Write your questions down as they come up or you might forget them!

# A Quick Note about Maths

There are five formulæ given on the formulæ sheet that you are expected to use. They are given in letter form, eg. E=m×c×ΔT. You do, however, need to know what each letter stands for, so here they are in words:

- $energy\ required = mass \times shc \times change\ in\ temperature$

      (J)             (kg)   (J/kg/°C)        ( °C)

  where shc= specific heat capacity

- $$efficiency = \frac{useful\ energy\ out}{total\ energy\ in}$$

  OR $$efficiency = \frac{useful\ power\ out}{total\ power\ in}$$

- $energy\ (kWh) = power(kW) \times time(h)$

  OR $\ energy\ (kWh) = power(kW) \times time(h)$

- $cost(p) = energy(kWh) \times cost\ per\ kWh(p/kWh)$

- $wavespeed\ (m/s) = frequency(Hz) \times wavelength(m)$

Here are a couple of other formulæ that are likely to come in handy:

- $total\ energy = useful\ work\ done + wasted\ energy$

- $$payback\ time\ (yrs) = \frac{initial\ cost\ (£)}{annual\ saving\ (£/yr)}$$

Foundation tier papers always give the numbers in such a way that the formulæ can be used in this form. Higher tier papers are likely to need them rearranging before use – remember how to rearrange by doing the same thing to both sides of the equation.

Higher tier candidates also have to watch units. Some values will not be given in the correct units (eg. cm instead of m) and will need to be converted. Foundation tier candidates are spared this extra step!

# A Quick Note about How Science Works

Scientists look at evidence. This comes from experiments and observations of the world. They then devise models or theories about what causes these things to happen. A good theory fits the current observations AND makes new predictions that can be tested.

Old theories get replaced by newer ones IF the new one is simpler or if it fits new evidence that the old one doesn't.

If you are asked to carry out an experiment to test a theory, you need to consider the following things:

- Is your sample size large enough?

  If not you may just be seeing a random effect. If three out of four tests of a new car tyre showed it reduced the braking distance, they might just be three sets of tyres that were particularly well made.

- Have you eliminated the effects of other factors?

  If not they could be affecting your results. Perhaps the tyres don't work well on wet roads.

- Have you eliminated any bias?

  If not you might not accept the evidence. If the company making the tyres is paying you to test their product, you might be worried about not getting hired again if you say it doesn't work.

A classic example of poor evidence is that given in shampoo adverts. For instance one claims 90% of those asked said their hair was silkier with the brand. Initially that sounds impressive.

Looking at the small print you see that the company itself asked 30 of their repeat customers. This is far too small a sample and is also biased.

In reality, it is sometimes unreasonable to carry out an experiment. This could be because:

- It would be too expensive.
- It would be unethical, for instance an experiment that could harm people.

# P1.1.1 Infrared Radiation

*Infrared radiation carries heat energy long distances without the need to anything to carry it. It even travels through the vacuum of space, unlike any other method of heat transfer. Without infrared radiation, we wouldn't be able to feel the Sun's heat.*

One of the ways that heat can be transferred is by infrared radiation.

- All objects emit infrared radiation. Even cold objects emit some.
- The hotter an object gets, the more infrared radiation it emits every second.

Different surfaces emit and absorb infrared radiation to different degrees.

- Dark, matt surfaces are particularly good at emitting infrared radiation. (Matt is the opposite of shiny. You might also see it called 'dull'.)
- Dark, matt surfaces are also good at absorbing infrared radiation.

- Shiny, light surfaces are particularly poor emitters of infrared radiation.
- They are also bad at absorbing it. That is because they reflect it away instead.

Black is a good absorber but also a good emitter. People often get stuck because they think that the heat can only go one way.

In Mediterranean countries, traditional male clothing is often white whilst traditional female clothing it often black. This is because the men tended to work in the fields and needed clothes that reflected the infrared radiation from the Sun to keep them cool. The women tended to work at home in the shade, and dark clothing was good at emitting infrared radiation to keep them cool.

Many sports, like tennis and cricket, have a tradition of wearing white for the same reason.

# P1.1.2 Kinetic Theory

*Kinetic theory explains most things about how materials behave. For a long time it was simply a model; there was no proof it was actually what was happening. It was very good at explaining and predicting behaviour and so scientists used it. Eventually, more and more direct evidence for the existence of the particles was found, and it became accepted that did actually describe how the materials were constructed.*

Particles can be used to describe and explain the behaviour of different states of matter.

- Solids have the particles organised in a regular pattern, like squares or hexagons.
- The particles in a solid are touching each other.
- The particles in a solid vibrate but don't move around.
- This means that a solid holds its shape.

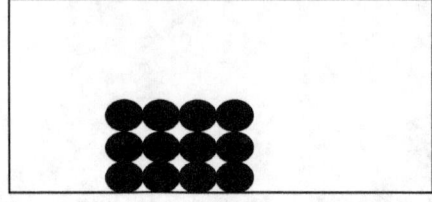

- Liquids don't have the particles in a regular pattern.
- Most of the particles in a liquid are touching each other.
- The particles in a liquid move past each other slowly.
- This means that a liquid flows to fill the bottom of its container.

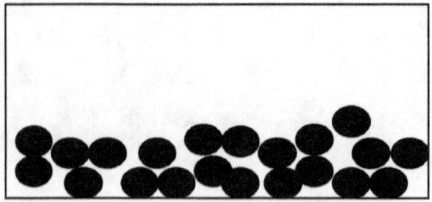

- Gases do not have the particles in a regular pattern.
- The particles in a gas are widely spread out.
- The particles in a gas move around rapidly and randomly.
- This means that a gas expands to fill its container.

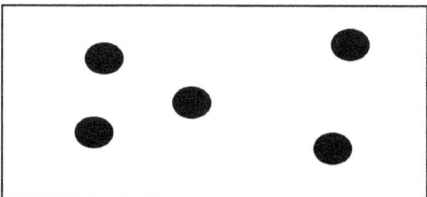

If you take a fixed mass of a substance and heat it from solid to liquid to gas, you will find that the energy in each state is higher than the one before, because the particles are moving faster.

# P1.1.3 Energy Transfer by Heating

*As well as infrared radiation, there are three other mechanisms for heat transfer. These apply in many everyday situations, including cooking, heating a house and keeping cool during exercise.*

Conduction is a heat transfer mechanism that involves the passing of heat energy from one particle to the next.
- Conduction works best in solids.
- Gases are very good insulators. Most insulating clothing works by trapping air.
- In a solid insulator, the vibrations of 'hot' particles are passed on from one to the next by the particles bumping into each other.
- Some examples of solid insulators are: wood, glass, rubber and plastic.
- In a solid conductor, heat energy is mostly transferred by the free electrons moving around and hitting particles in the solid. That is why heat moves faster through a conductor than an insulator.
- Metals are good conductors because they have lots of free electrons.

Convection is a heat transfer mechanism that works in liquids and gases (and solids under immense pressure, but you don't need to worry about that).
- Solids and liquids can be called fluids.
- When one region of a fluid is heated, the particles there move further apart. It expands.
- That means it becomes less dense than its surroundings, and so rises to the top.
- On reaching the top (or cooling so much that it stops rising) it is pushed sideways by the fluid coming up from below.
- It is then able to sink back down further away once it has cooled, contracted and become more dense again.

Evaporation is another heat transfer mechanism.
- In a liquid the particles take a range of energies.
- The most energetic particles can escape. These particles form a gas.
- The particles with less energy are left behind, so the liquid is now cooler than before.

If a gas is cooled the particles lose energy and can turn into a liquid.
- This is called condensation.

The rate of evaporation and of condensation can be increased by:
- increasing the surface area that is exposed to the air
- increasing the temperature difference between the surface and the air
- using a more volatile liquid, ie. one which evaporates more easily

Remember that radiation is another heat transfer mechanism (see P1.1.1).

There are several things which affect the rate that heat energy flows out of or into an object.
- The higher the surface-area-to-volume ratio of an object, the faster heat will be transferred.
- The larger the temperature difference between an object and its surroundings, the faster it will transfer heat.
- The rate of heat flow also depends on what the object is made of and what it is in contact with.

## P1.1.4 Heating and Insulating Buildings

*Some things you want to keep hot, like your tea on a long walk. Other things you want to keep cold, like your ice cream. Unless you insulate them, heat will flow into or out of them until they end up the same temperature as their surroundings.*

To reduce the amount of heat escaping from a building we use insulation.
Examples include:

- cavity wall insulation
- loft insulation
- carpets
- double glazing
- draught excluders.

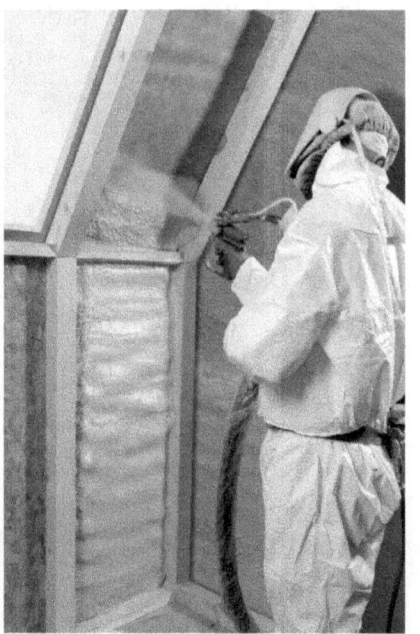

The effectiveness of an insulating material is called its U Value.

- The U Value shows the rate at which heat flows through the material.
- The lower the U Value, the better insulator the material is.

Most heating systems for buildings involve heating a liquid (eg. oil or water) or a solid (eg. ceramic bricks or metal). Designers need to be able to predict how much energy is needed to heat a material up by a certain amount.

- The specific heat capacity of a material is how much energy is needed to heat 1kg of it by 1°C.

$$energy\ required = mass \times shc \times change\ in\ temperature$$
$$\quad (J) \qquad\qquad (kg) \quad (J/kg/°C) \qquad\qquad (°C)$$

$$E = m \times c \times \theta$$

Not all buildings are heated using central heating powered by fossil fuels or electricity. Some central heating systems use solar water heating. In hot countries, this can completely replace the traditional boiler; in colder countries it is used as a top-up to reduce their bills and carbon footprint.

- Installing solar water heating involves putting panels facing the sun. These are often put on the roof.
- The panels are painted black to absorb the most heat radiation possible.
- There is often also a silvered layer behind to reflect any infrared that got through back onto the panel.
- Inside them are pipes through which water is pumped. The water is heated by the heat captured from the sun.

When a house is sold nowadays, it is required to have an energy performance certificate. This indicates how energy efficient it is and includes information on insulation and other energy reduction methods.

| Energy Efficiency Rating | | |
| --- | --- | --- |
| | Current | Potential |
| Very energy efficient - lower running costs | | |
| (92-100) A | | |
| (81-91) B | | |
| (69-80) C | | 70 |
| (55-68) D | | |
| (39-54) E | 52 | |
| (21-38) F | | |
| (1-20) G | | |
| Not energy efficient - higher running costs | | |
| UK 2005 | Directive 2002/91/EC | |

# P1.2.1 Energy Transfers and Efficiency

*Energy rarely stays in one place. It keeps changing its form, and transferring from one object to another. Sadly these changes are never 100% efficient and energy appears to be lost along the way.*

One of the most important and fundamental principles of the universe is the Conservation of Energy.
- The Principle of Conservation of Energy states that energy cannot be created or destroyed.
- Energy can, however, be changed from one form to another. Eg. from kinetic to heat in a brake.
- Energy can also be passed from one object to another. Eg. from bat to ball.

These transformations can lead to the energy being stored, like the chemical energy in a battery.
They can also lead to useful work being done, like kinetic energy in a car.

However, the transfers and transformations also tend to lead to energy being spread out so it cannot be retrieved, like the heat from an electrical wire.
- We call this dissipation.
- Once energy has been dissipated, it ceases to be useful - we say it is wasted.
- One common form this wasted energy takes is heat energy in the surroundings. Another is sound energy.

Efficiency is a measure of how much of the total energy available is usefully used.
- We can calculate the efficiency of a transformation or transfer:
$$efficiency = \frac{useful\ energy\ out}{total\ energy\ in}$$
- Alternatively,
$$efficiency = \frac{useful\ power\ out}{total\ power\ in}$$

The formulae opposite give you a decimal or fraction, depending on your calculator.
- The number should always be less than 1.
- If you are asked for a percentage, you need to multiply the number by 100.

Remember, energy isn't actually lost.
- The total power that is input is always equal to the total power that is output.
- If you know them, you can add up all the outputs to work out the input.

Energy transformations and transfers are often represented using Sankey Diagrams (also known as energy transfer diagrams).
- These show the input energy and the output energies in pictorial form.
- The width of each arrow shaft represents the amount of each form of energy.
- The width of each arrow shaft on the right will add up to the width of the arrow shaft on the left.

Wasting energy can be a big problem. It can increase fuel and electricity costs. It also increases the use of fossil fuels, which are running out.
- Some devices are designed to minimise the amount of energy wasted, eg. compact fluorescent tube and LED light bulbs give off less heat as traditional filament bulbs, whilst emitting the same amount of light.
- Insulating your house is another example of a way to reduce energy wastage.
- There are also schemes available that capture the waste energy and use it again, eg. using waste hot water from a power station to heat buildings or energy from a crematorium to heat a public swimming pool.

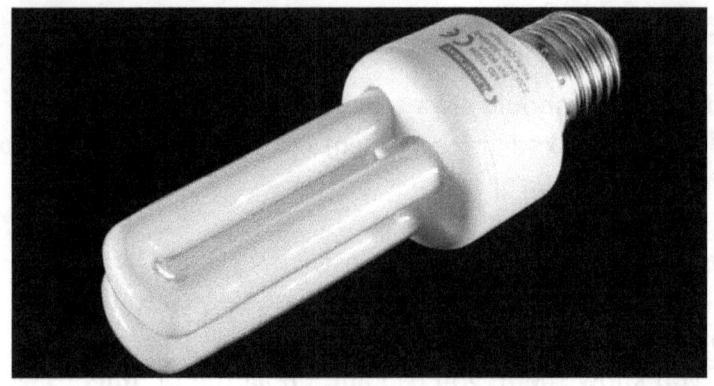

Most things you could do to improve energy efficiency will cost you money to install.

- The more energy efficient a house it, the less money needs to be spent on bills.
- There is usually a capital cost paid up front to buy the equipment and install it.
- A common way to judge if it is worth installing something as an energy saving measure is to calculate its payback time. The payback time is how long it would take to reclaim your capital cost from the savings in your energy bills.
- The cost-effectiveness of an energy saving measure can also be judged on how much money you will save in the lifetime of the equipment. The larger the saving you make in that time, the more cost-effective the measure is.

Eg.  A new boiler costs £2000 and lasts for 18 years. You save £200 a year on your heating bills.

This means your payback time is 2000/200=10 years.

In its lifetime it will save you 18x£200 - £2000 = £1600.

Fitting a draught excluder costs £6 and lasts for 25 years. It saves £3 year on your heating bills.

This means your payback time is 6/3 = 2 years.

In its lifetime it will save you 25x£2 - £6 = £44.

The draught excluder has a shorter payback time but the new boiler is more cost-effective in the long-run.

You might be asked to compare the amount of money saved by different measures in a given number of years.

- The one which saves the most money in that time is the best.
- Sometimes the amount saved is negative - that means you haven't yet paid back the initial cost.

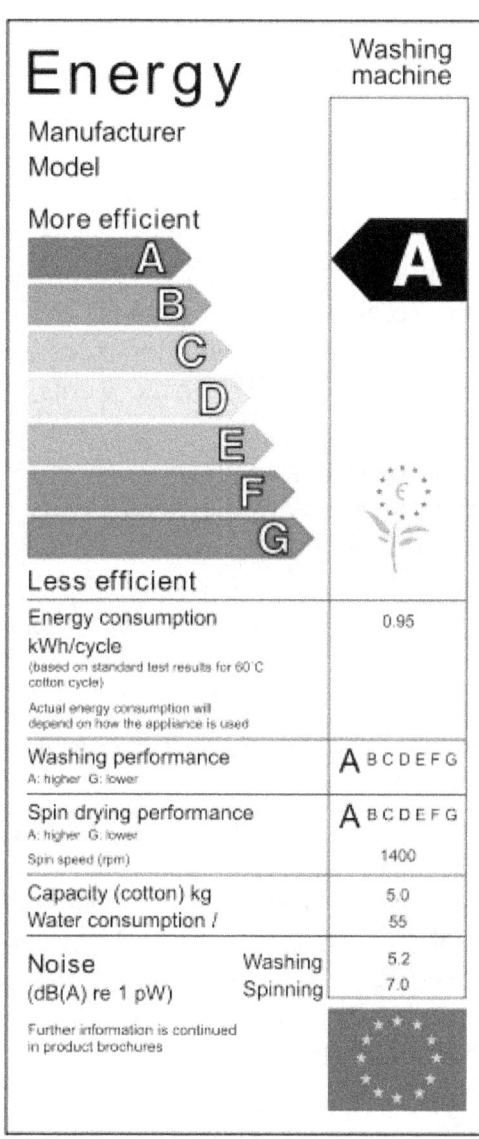

## P1.3.1 Transferring Electrical Energy

*Most devices change one form of energy into another. Nowadays, most domestic appliances run on electricity - they convert electrical energy into something else.*

There are many forms of output energy.
- If the output is movement, then the output energy is kinetic.
- Other examples include sound, heat and light.

It is often useful to be able to predict how much energy a device is going to use. This could help you decide whether it is worth buying, or just help to budget your utility bills. Some have an energy performance rating, but others don't – you will have to do your own calculations to compare those that don't.
- The higher the power of the device the more energy is used.
- The longer the device is on for, the more energy is used.
- The formula for this is:

$$energy\ (J) = power(W) \times time(s) \qquad E = P \times t$$
- However, when dealing with the electricity used in the home it is easier to use different units:

$$energy\ (kWh) = power(kW) \times time(h)$$
- kWh stands for kilowatt-hour – the amount of energy used if a 1kW device is on for 1 hour.

You can see how much electricity you are using by reading your electricity meter. Some even come with an app for your 'phone!
- The electricity meter tells you how many kWh have been used since it was installed.
- To find out how much you used in a certain time, note down the reading at the start then subtract that number from the reading at the end.

Eg. Last Monday the reading was 58123. This Monday it is 58195.

   We used 58195-58123=72kWh in the last week.

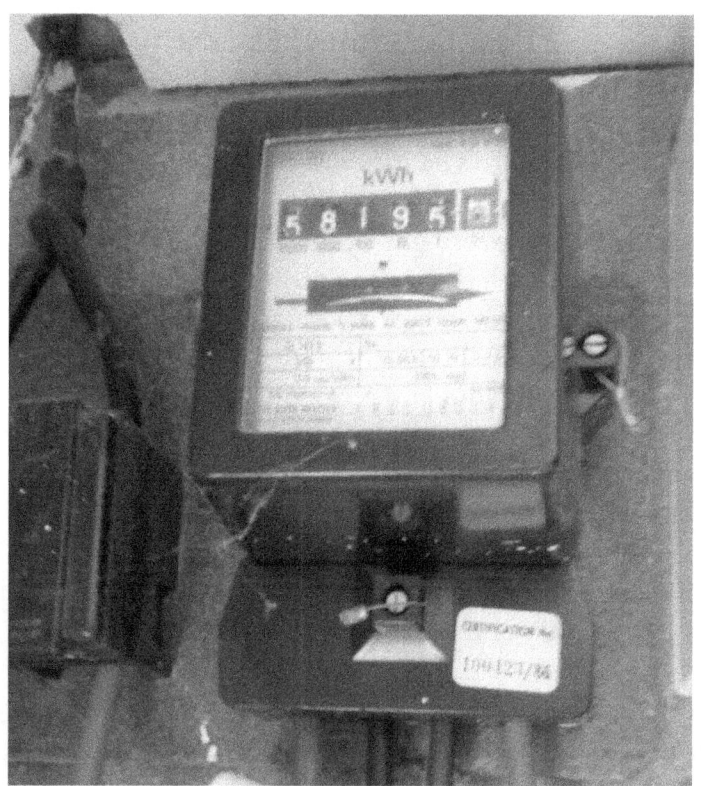

Calculating the cost of the electricity is really easy.

- The power companies charge you for every kWh of energy you use.
- You can find this on your last bill, or by logging in to your online account with the company.
- You can use the following formula to work out how much you will be charged:

$$cost(p) = energy(kWh) \times cost\ per\ kWh(p/kWh)$$

Eg. We used 72kWh last week, and the company charges us 10.5p per kWh.

72×10.5=756p

We used £7.56 worth of electricity in the last week.

In addition, the companies also add a standing charge to your bill, for maintenance of your supply.

# P1.4.1 Generating Electricity

*We use a lot of electricity. It is generated for us in a large variety of ways, some will keep working forever, others will run out soon. They all have different impacts on the environment.*

Some energy resources are non-renewable.
- These include coal, oil, gas and nuclear.
- Non-renewable resources have very limited amounts left on Earth.
- They are likely run out in the next hundred or so years.
- Some might run out in your lifetime.

Other energy resources are renewable.
- These include wind, tidal, hydroelectric, geothermal, wave, biomass and solar.
- Renewable resources are replenished within a human lifetime and so will not run out on us.

The most common form of power station involves boiling water to make steam.
- This steam moves under high pressure down a pipe and turns a turbine.
- The turbine spins a generator, which makes electricity.
- The steam is then cooled and condensed.
- The resulting water is piped back into the boiler to complete the cycle.
- Fossil fuel power stations use fuels such as coal, oil and gas as the energy source.
- It is also possible to use renewable biofuels, like woodchip or methane from animal dung and rotting plant matter.

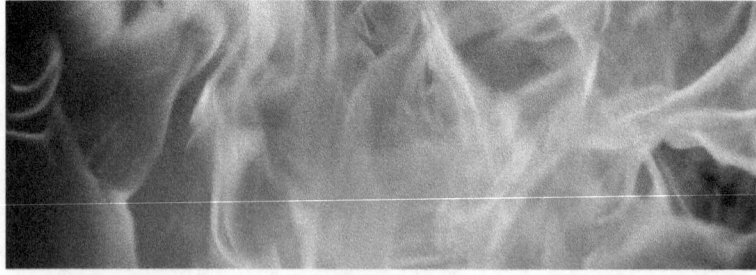

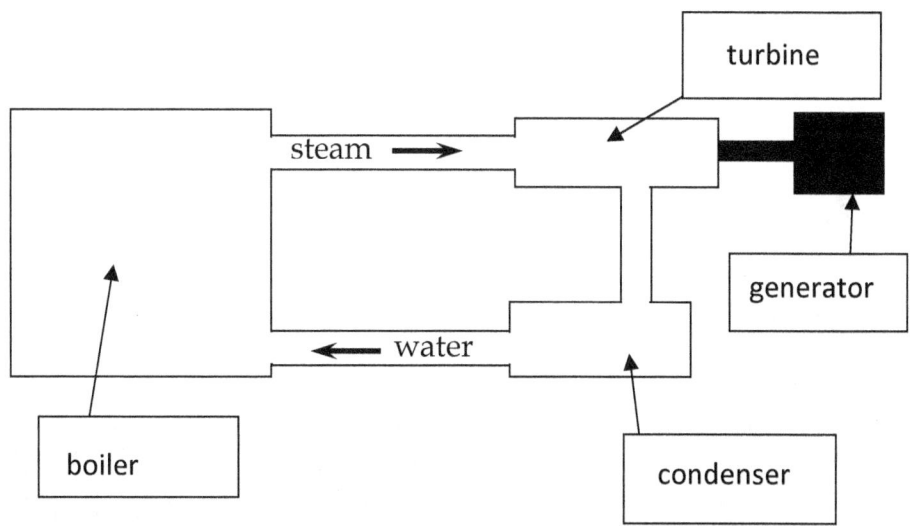

A very similar process occurs in a nuclear power station.

- Nuclear power stations use uranium or plutonium as fuel.
- The fuel doesn't burn, it undergoes a process called fission.
- Fission releases heat.
- After that, the process is the same as for a fossil fuel power station – heat boils water, steam turns turbine etc.

In some regions there are very hot rocks near the surface.
- Companies drill down and pump water into the ground.
- Hot water, or even steam, comes back out.
- This can be used to heat homes or make electricity.
- This is called geothermal energy.

In sunny countries, the Sun's heat can be used to make electricity.
- Solar towers use mirrors to focus the Sun's energy onto pipes of water at the top of the tower.
- These pipes are black to absorb as much infrared radiation as possible.
- The temperature can reach hundreds of degrees and the heat energy is used to boil water.
- The steam can then be fed into a traditional turbine system.

There are ways of generating electricity that don't involve boiling water.

- Wind farms use the kinetic energy of the wind to turn the turbine directly.
- Hydroelectric power stations use the kinetic energy in moving water, usually released from a dam, to turn the turbines.
- Tidal barrages trap the water at high tide and release it through turbines at low tide.
- Tidal turbines are like wind farms underwater, catching the flow of water through a gap between islands.
- Most solar farms use photovoltaic cells, like on a calculator, to convert sunlight directly into electricity.

There are several things that need to be considered when choosing between resources.

- Does it produce harmful waste?

    Fossil fuels release carbon dioxide ($CO_2$) and sulphur dioxide ($SO_2$) into the atmosphere. $CO_2$ causes global warming and $SO_2$ causes acid rain.

    Nuclear power stations produce radioactive and toxic waste that needs to be transported and stored safely for thousands of years.

- Is it reliable?

    Solar only works in the daytime. It isn't as effective on cloudy days.

    Tidal works twice a day every day. It is predictable, but not always when you need it.

    Wind only works if it is windy. For safety reasons, wind farms have to be switched off if it is too windy.

- Does it damage animal and plant habitats?

    Hydroelectric often requires valleys to be flooded.

    Wind turbines are accused of killing birds.

    Tidal can destroy the estuary environment.

- Will it offend the neighbours?

    A lot of people object to the look and sound of wind farms.

    There is a strong opposition to nuclear power on perceived safety grounds.

- How quickly can it start generating when electricity is needed?

    Power stations that involve boiling water can take hours to get going - they can't react to sudden surges in demand like everyone making a cup of tea during half-time in a match.

    Of the fossil fuel power stations, gas is the fastest, but it still takes 15 minutes or so to start producing electricity.

    Pumped storage is a good way to react to surges – water is pumped up a mountain to a lake, using spare electricity when it is available, then the water is released to generate electricity if a surge in demand is detected. This can work in seconds and is often automated.

- How much will it cost?

   Some systems are cheap and quick to build.

   Others require lots of land to be purchased and massive infrastructure to be constructed.

      You need to include decommissioning costs, particularly those involved in making nuclear power stations safe for future generations.

- Is it suitable for the location?

      Building a coal-fired power station on a small island would require millions of tons of coal to be ferried out, so something that doesn't need fuel, like tidal or wind, would be better.

      Small-scale applications, like illuminated road signs and lights, often use solar cells or small wind turbines.

Despite years of campaigning, research and government funding, we are still heavily reliant on fossil fuels.

- This means that the release into the atmosphere of $CO_2$ is still a big problem.

- One way to reduce this effect is to plant more trees. These will absorb some $CO_2$ through photosynthesis.

- Another way to reduce the amount getting into the atmosphere is artificial 'carbon capture'. The $CO_2$ is trapped before it leaves the power station and pumped into storage. The best candidates so far for the storage location are disused oil and gas fields, like the ones under the North Sea.

- New fossil fuel power stations must now come with a plan to limit their carbon emissions.

## P1.4.2 The National Grid

*We can generate lots of electricity, but without some sort of supply network it won't get where it is needed. Once it is generated, it has to be delivered to our homes and businesses. That is what the National Grid is for.*

The National Grid is a system of cables, transformers and pylons which transfer electricity around the country.

- The National Grid links the majority of power stations to the majority of businesses and homes.
- It allows supply and demand to be balanced across the country – if Liverpool is using an excessive amount of electricity, for instance during a big local football match, a power station in Somerset could cover it.
- If one power station needs to be repaired, the others around the country can cover for it.
- The power stations and customers are not part of the National Grid, they are just connected by it.

The electricity is transported at very high voltages.

- This means that the current is much smaller than produced y the power station and so the cables heat up less.
- The less energy that is wasted as heat, the more electrical energy gets to the users.
- In order to step the voltage up for transmission, and down again for safer use, the National Grid uses transformers.

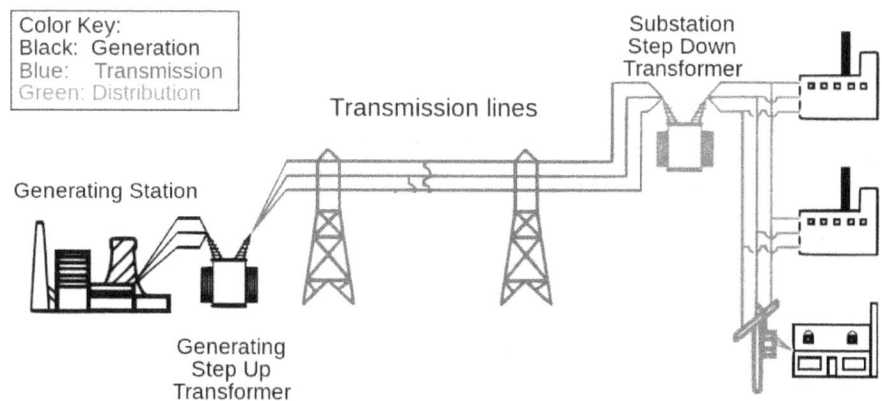

Color Key:
Black: Generation
Blue: Transmission
Green: Distribution

Transmission lines

Substation
Step Down
Transformer

Generating Station

Generating
Step Up
Transformer

Across most of the country, the electricity is carried in overhead power lines. In some places they are buried underground.

- Overhead power lines are easier to get at for maintenance, cheaper to install and don't carry the risk of someone digging into them or getting flooded.
- Underground power lines avoid the problems of looking ugly and being hit by low-flying aircraft.

Not everyone is a simple receiver of power from the National Grid.

- Some homes are not connected, either because they are too remote or their owners don't want to pay for electricity.
- Some homes have their own power generators, eg. solar panels. The National Grid buys their surplus electricity when they have it, and supplies them when they can't make enough themselves.

## P1.5.1 General Properties of Waves

*Waves are everywhere. They are used for a lot of purposes, one in particular being for communication. One class of waves that are really important to us are electromagnetic waves.*

Waves are formed when vibrations in a medium get passed on.
- They carry energy, but do not move the matter from one place to another.
- They can carry information.

Transverse waves are one of the two classes of wave.
- In transverse waves, the vibrations (also known as oscillations) are perpendicular to the direction of travel of the wave.
- They have peaks (high points) and troughs (low points).
- Electromagnetic waves, such as infrared, radio and visible light, are examples of transverse waves.

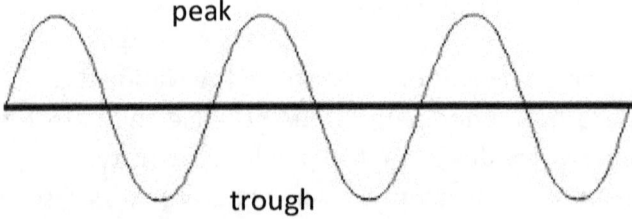

peak

trough

The other class is called longitudinal.
- The vibrations are parallel to the direction of travel of the wave.

- They have compressions and rarefactions instead of peaks and troughs.

compression

rarefaction

- Compressions are where the particles are close together.
- Rarefactions are where the particles are far apart.

- Sound waves and earthquake p-waves are examples of longitudinal waves.

Mechanical waves are those which actually move physical particles back and forth.
- Mechanical waves can be either longitudinal or transverse (or a combination of the two).
- All waves, apart from electromagnetic waves, are mechanical.

There are several key properties of waves that you need to know.
- Wavelength is the distance between two consecutive, identical parts of the wave. It is easiest to measure peak to peak, but trough to trough or riser to riser is fine too.
- Frequency is how many complete waves pass a point in 1 second. It is easiest to count peaks.
- Amplitude is the distance from the undisturbed position to the top of a peak. It can sometimes be easiest to measure from trough to peak and divide by two.

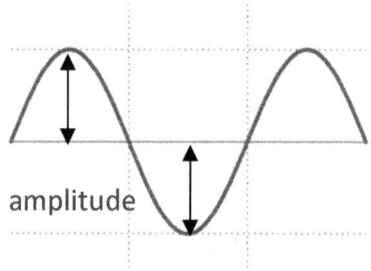

amplitude

All waves obey the wave equation.

$$wavespeed \ (m/s) = frequency(Hz) \times wavelength(m)$$
$$v = f \times \lambda$$

All waves can have their directions of travel changed.
- This could be by reflection, refraction or diffraction.

Reflection is dealt with in the next chapter (P1.5.2).

Refraction is when a wave changes direction because it passes across a boundary between two different media.

- Refraction does not occur if the wave hits the boundary at right angles, ie. along the normal line.

Diffraction is when a wave spreads out on passing through a gap.

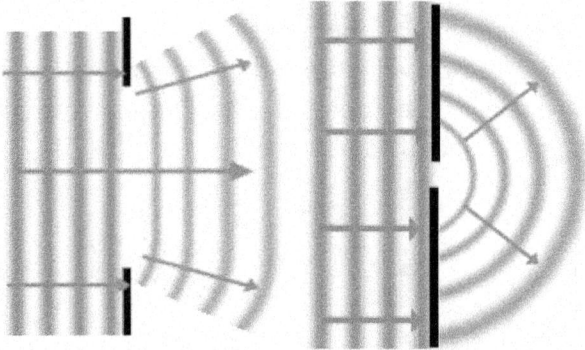

- The most diffraction occurs when the gap size is similar to the wavelength of the wave.
- Diffraction also occurs around obstacles.
- You can sometimes see waves on the sea diffract round breakwaters.

All electromagnetic (EM) waves travel at the same speed in a vacuum.
- They travel at the speed of light: $3 \times 10^8$ m/s.
- The electromagnetic spectrum is continuous but we divide it into sections to make it easier to talk about.

Starting with the longest wavelength on the left, these sections are: radio, microwave, infrared, visible, ultraviolet, x-rays, gamma rays

- Radio has a wavelength of about 10,000m. Gamma rays have a wavelength of about $10^{-15}$m.
- The order above is also that of decreasing frequency.
- It also shows the order of increasing energy.
- The further right you go on the spectrum above, the more harm the waves do to you.

Waves are used in many ways for communications.
- Sound is used for speech and things like referees' whistles and car horns.
- Radio waves are used for radios and televisions.
- Microwaves are used for mobile 'phones and satellite communications.
- Infrared is used for television remote controls and some fibre-optic applications, eg. broadband internet.
- Visible light is used for things like photography, flashing morse code, semaphore and hand gestures.

# P1.5.2 Reflection

*We are all familiar with reflections. But explaining how they form can be a little tricky!*

Scientists use ray diagrams to show how light travels.
- Rays are straight lines because light travels in straight lines.
- Rays have arrows on them to show the direction of travel of the light.
- Often angles need to be measured. These are always measured from the 'normal line'.
- The 'normal line', often just called the 'normal', is drawn at right angles to a surface or boundary at the point where a ray hits it.

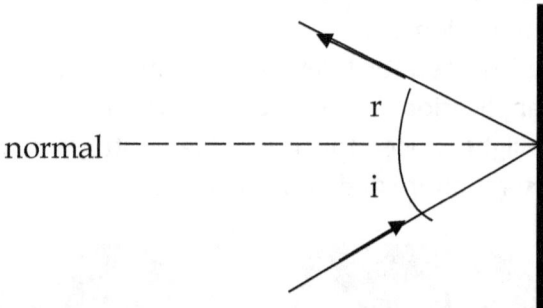

The law of reflection states that 'the angle of incidence is equal to the angle of reflection'.
- In the diagram above, that means that i=r.

When you look at yourself in a mirror, you see the image of your face the right way up but with left and right swapped round.
- We say that the image is 'laterally inverted'.
- Another thing to note is that it is a virtual image. That means that it appears to be formed behind the mirror, rather than being projected onto a wall or screen.
- A virtual image is actually formed when our brain tries to interpret the rays hitting your eyes.

- If you are looking in a flat (also known as 'plane') mirror, then your image will be the same distance behind the mirror as you are in front of it.

We can use ray diagrams to explain why the image is formed where it is.
- Firstly, choose the point on the object you want to work out.
- Draw a couple of rays from that point to the mirror.
- Next, draw the reflected ray for each one. Remember to obey the law of reflection.
- Now, the brain assumes the light has travelled in straight lines to get to it. This means it traces the rays back to a point. Do the same with your reflected rays.

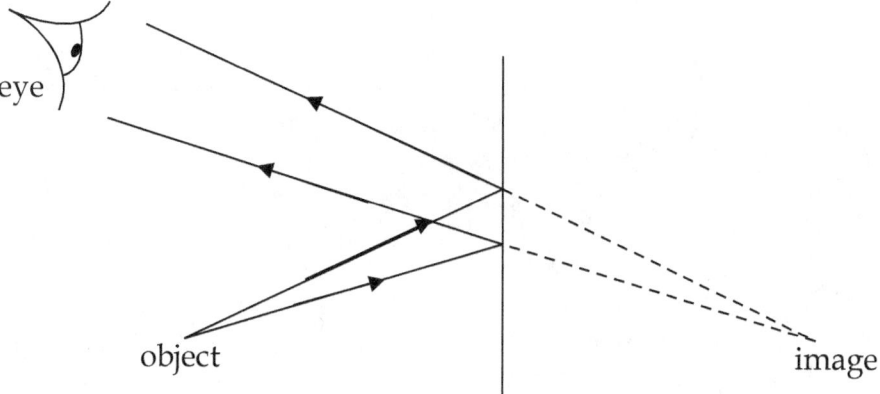

- You can repeat this for other points to form the entire image.

# P1.5.3 Sound

*We communicate through sound when we speak. We also use it for warnings, like with a car horn, and when listening to music.*

Sound waves are caused by objects vibrating.

- These vibrations cause the air around the object to be pushed back and forth.
- This causes regions of high and low pressure to be formed.
- These regions are called compressions and rarefactions respectively.
- The vibrations of the air particle are parallel to the direction of travel of the wave.
- This means that sound is a longitudinal wave.
- Sound can also be transferred by solids and liquids.

What the sound sounds like is determined by how the object vibrates.
- The higher the frequency of the vibrations, the higher the note that is heard. It is higher pitched.
- The larger the amplitude of the vibrations, the greater the volume of the note that is heard. It is louder.

Sound waves can bounces off objects.
- We call this echoing.
- Echoes can be used to locate objects, like fish or submarines.
- This is called sonar.
- Some animals, such as dolphins and bats, use sonar to find their way around and hunt their prey.

## P1.5.4 Red Shift

*Electromagnetic waves are used by astronomers to observe stars and other objects in space. They can give us an insight into the very origins of the universe.*

If you are stood on a street and an ambulance goes past, the siren sounds high pitched as it approaches you and low pitched as it goes away again.

- This is an example of something called the Doppler Effect.
- When the source of a wave moves towards an observer, the wavelength they detect goes down and the frequency goes up.
- When the source of a wave moves away from an observer, the wavelength they detect goes up and the frequency goes down.

When applied to light, we use colours to describe the change.

- The process where the wavelength gets longer is called red shift.
- The process where the wavelength gets shorter is called blue shift.

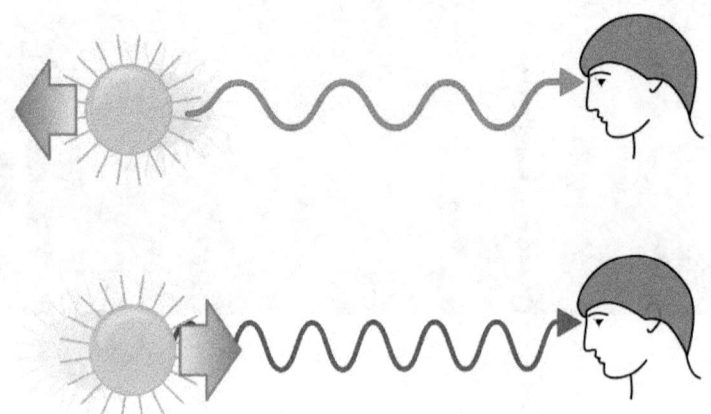

We can use an attachment on a telescope, called a spectrometer, to split the light from a star into its colours. The spectrum produced contains dark lines. The wavelengths of these lines are characteristic of particular elements in the stars.

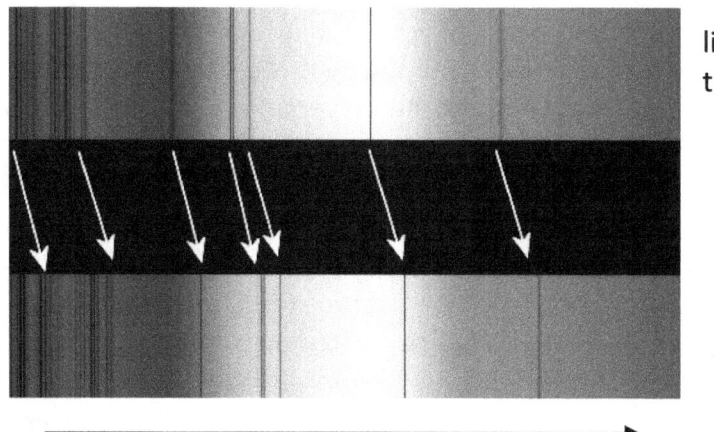

light from
the Sun

light from
the distant
star

increasing wavelength

- The position of the lines is compared to where we see them in the Sun.
- If they are further right on this diagram, then the light has been red-shifted.

When the light from stars in distant galaxies is studied, it is found to be red shifted.
- This shows that the distant galaxies are moving away from us.
- The further away the galaxy is, the more red-shifted the light from it is.
- We can conclude that the further away the galaxy is, the faster it is moving away from us.

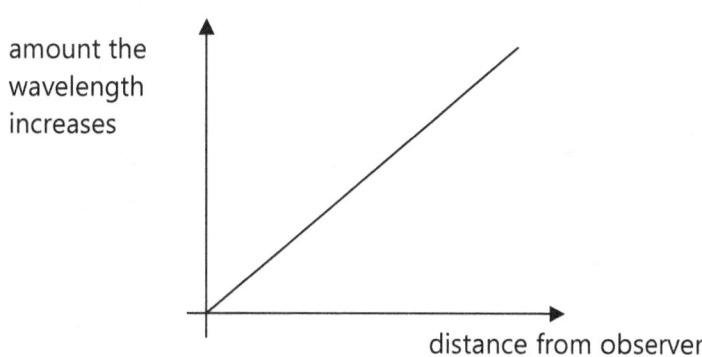

amount the
wavelength
increases

distance from observer

There are many theories about how the Universe was created.
- One model for the origin of the Universe is called the Big Bang.
- The Big Bang model says that the Universe was created in a tiny point and then expanded.
- Another model is the Steady State model.
- The Steady State model says that the Universe has always been the way it is. It has never changed.
- Most religions also have a creation story, such as Genesis, which usually have the Universe created by a deity.

When scientists decide between theories, they first examine whether they fit the evidence from experiment.
- The red shift of the distant galaxies provides evidence that the Universe is expanding. This supports the Big Bang model.
- The fact that the further away galaxies move faster is exactly what would happen if the universe were expanding from a tiny point.
- Another observation that supports the Big Bang model is Cosmic Microwave Background Radiation (CMBR). This is a weak microwave signal that can be observed in all directions and fits being radiation created in the Big Bang. It is sometimes referred to as an 'echo' of the Big Bang.
- Whilst some other models include the Universe expanding, none yet can explain CMBR.

# Also by the Author

**Physics**
Physics Problems for GCSE
http://www.archaeoroutes.co.uk/edphys/problems.php

The Best Bits of Physics
http://www.archaeoroutes.co.uk/bbop

AQA GCSE Physics Revision Guides
http://www.archaeoroutes.co.uk/edphys/revision.php

AQA A-Level Physics Practice Tests
http://www.archaeoroutes.co.uk/edphys/exams.php


**Science Fiction**
Independence
http://lrd.to/xZJGd4yPXo


**Archaeology and Walking**
Walking Through the Past series
http://www.archaeoroutes.co.uk

# Credits

The following images are licensed under a Creative Commons Attribution ShareAlike Unported 3.0 license (https://creativecommons.org/licenses/by-sa/3.0/deed.en):

**p14** - https://commons.wikimedia.org/wiki/File:WALLTITE_spray_foam_insulation_being_applied.jpg

**p18** - https://commons.wikimedia.org/wiki/File:Energiesparlampe_01_retouched.jpg

**p29** - https://commons.wikimedia.org/wiki/File:Electricity_grid_simple-_North_America.svg

**p30** - https://commons.wikimedia.org/wiki/File:Molecule5.gif

**p31** - https://commons.wikimedia.org/wiki/File:Simple_sine_wave.svg

**p32** - *Theresa knott at the English language Wikipedia* https://commons.wikimedia.org/wiki/File:Water_ripples_Diffraction.png

**p38** - https://en.wikipedia.org/wiki/File:Redshift_blueshift.svg

**p39** - https://en.wikipedia.org/wiki/File:Redshift.svg

The following images are licenced under a Creative Commons Attribution ShareAlike Generic 2.5 license (https://creativecommons.org/licenses/by-sa/2.5/deed.en):

**p23** – *RaBoe/Wikipedia* https://commons.wikimedia.org/wiki/File:Peitz_kraftwerk_jaenschwalde_sommer_nah.jpg

The following images are licenced under a Creative Commons Attribution ShareAlike New Zealand 3.0 license (https://creativecommons.org/licenses/by-sa/3.0/nz/deed.en):

**p24** - https://commons.wikimedia.org/wiki/File:Wairakei_Geothermal_Power_Station-5838.jpg

All other images are copyright Alasdair C Shaw, have been released into the public domain by their author, or are available under a CC0 license.

www.ingramcontent.com/pod-product-compliance
Lightning Source LLC
Chambersburg PA
CBHW071547170526
45166CB00004B/1575